The Adventures of the STEAM Machine

Slide, Beads, Slide!

Dr. Maury Wills

MEWE
Lithonia, GA

Publisher:
MEWE, LLC
www.mewellc.com

The Adventures of the STEAM Machine Series
Slide, Beads, Slide!

ISBN: 979-8-9871970-6-6
Library of Congress Control Number: 2023914626

For Worldwide Distribution
Printed in the USA

This book belongs to:

_____

4

Mr. Lee informed his class that the Super Solvers would be visiting the school to teach them about math tools. Amy was excited because math was her favorite subject.

6

When the Super Solvers arrived, Dr. Wills, the driver of the STEAM Machine bus, greeted the students.

"Hello students," said Dr. Wills. "Today, we're going to talk about the abacus and other tools that help us count, add, subtract, and do other math problems.

"Wow, that sounds like fun!" Amy said, eager to learn.

Brianna, the Math Super Solver, explained that the abacus had rows of beads that represented different values in math.

11

Brianna showed the students how to hold the abacus and how to use their fingers to move the beads.

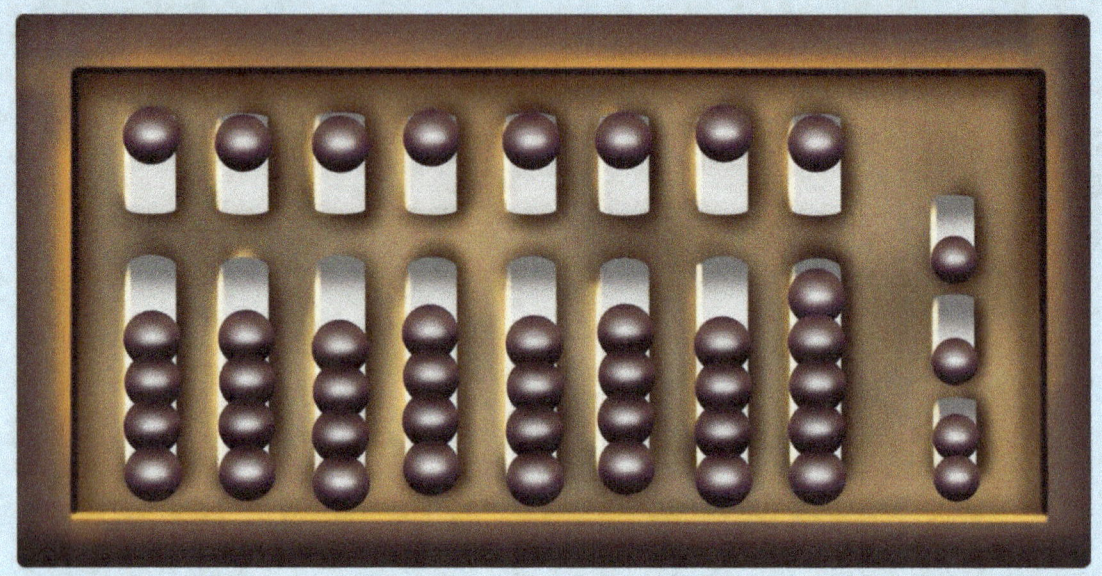

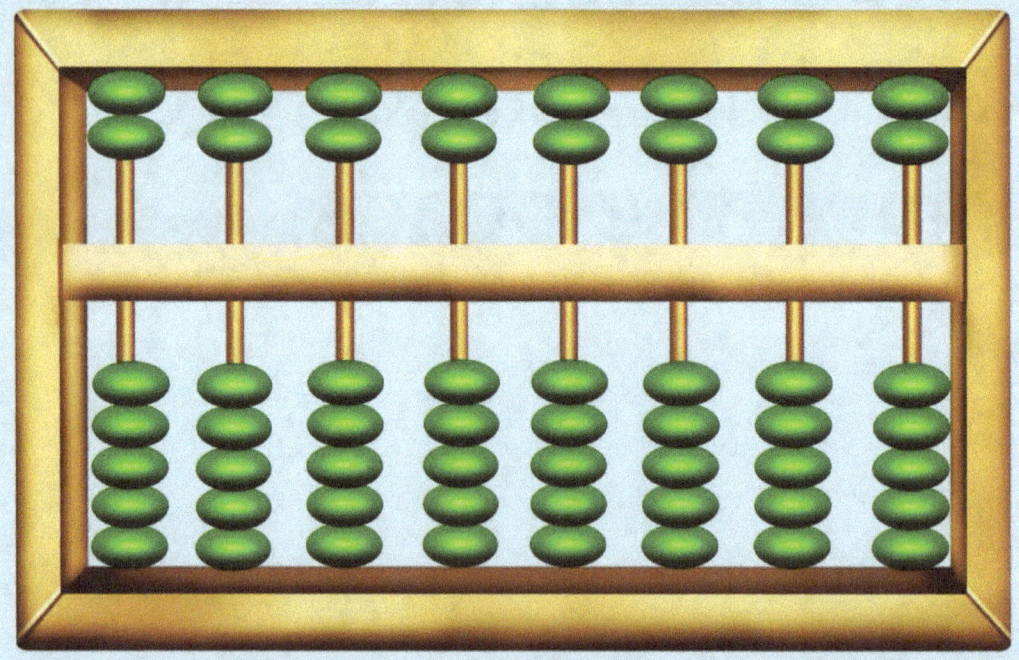

14

"The abacus was invented over 2,000 years ago," said Brianna. "It is still used in many parts of the world today."

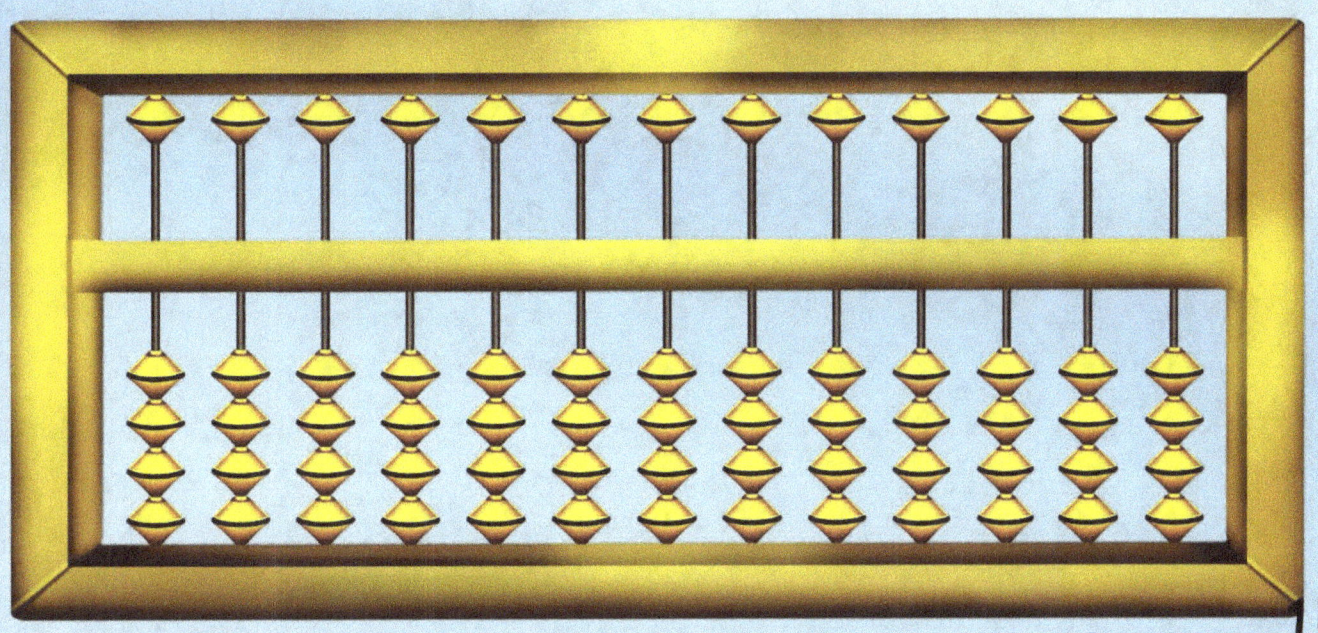

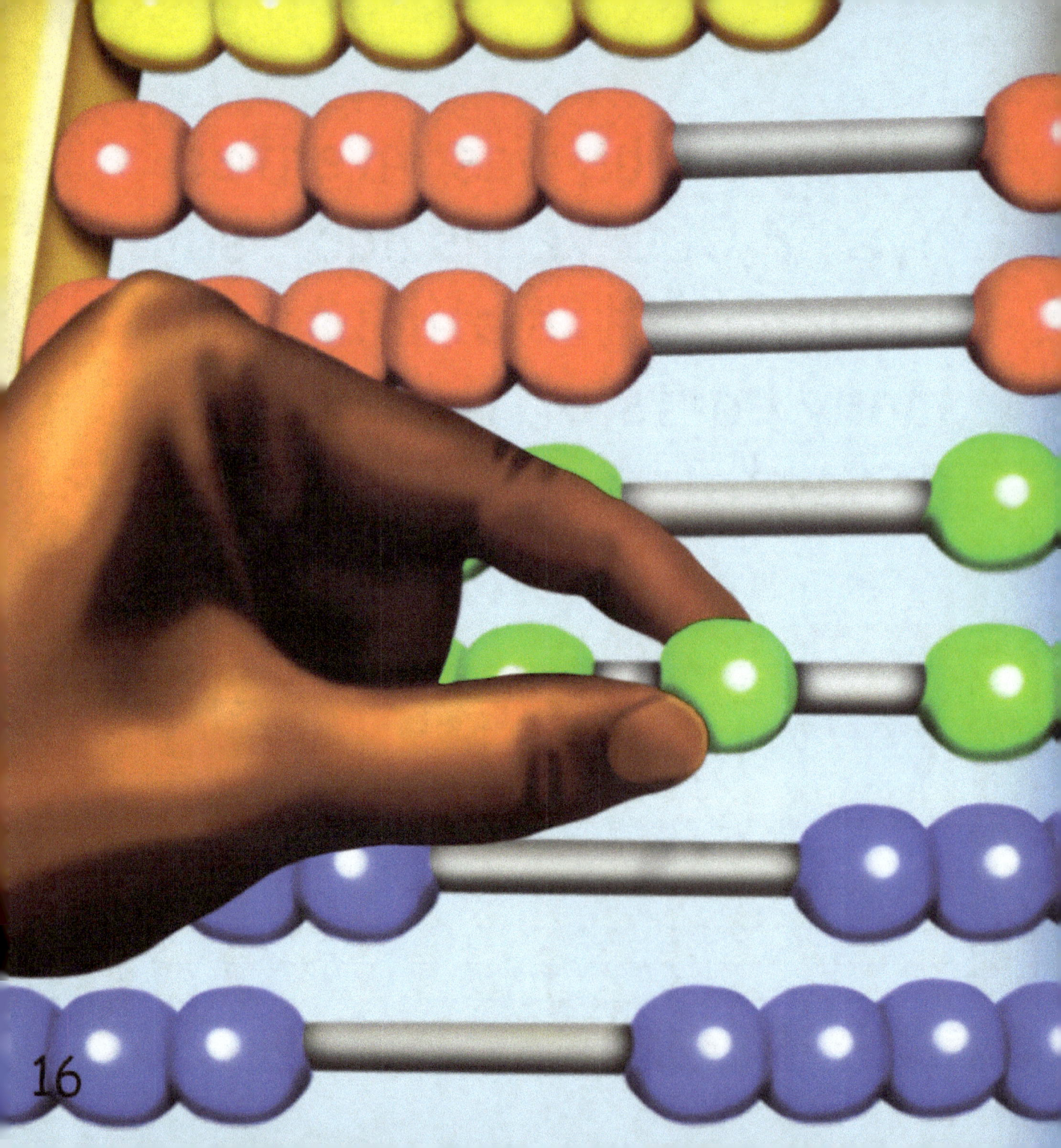

16

Brianna showed the students how to move the beads up to add and how to move the beads down to subtract.

18

Each student received an abacus to practice adding and subtracting.

19

Amy was amazed at how quickly she was able to do math using the abacus.

"Would you please show us some other math tools?" Amy asked.

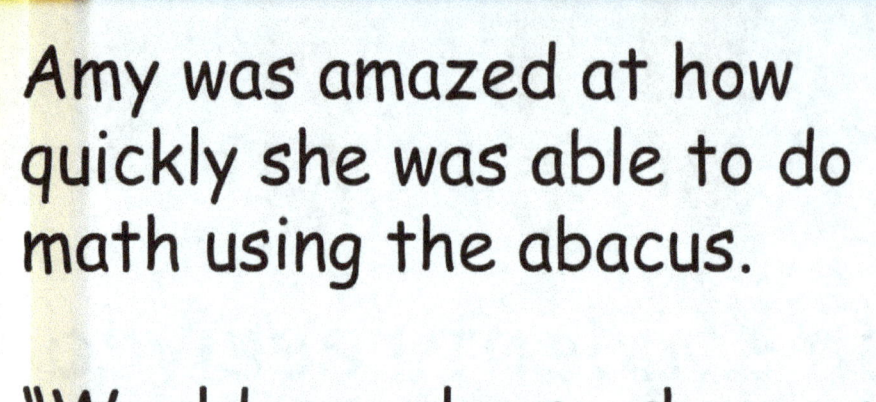

21

"Yes," Dr. Wills replied. "The slide rule was used by scientists before the calculator was invented, and the compass is still used by people who design buildings and create art."

Brianna showed the students a calculator.

"Calculators are great for doing math quickly," Brianna explained.

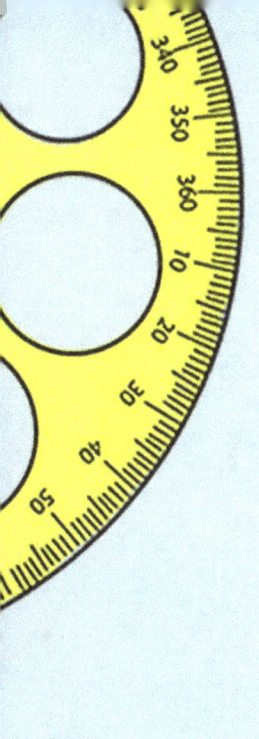

The students were excited to learn about the different math tools, and they couldn't wait to try all of them.

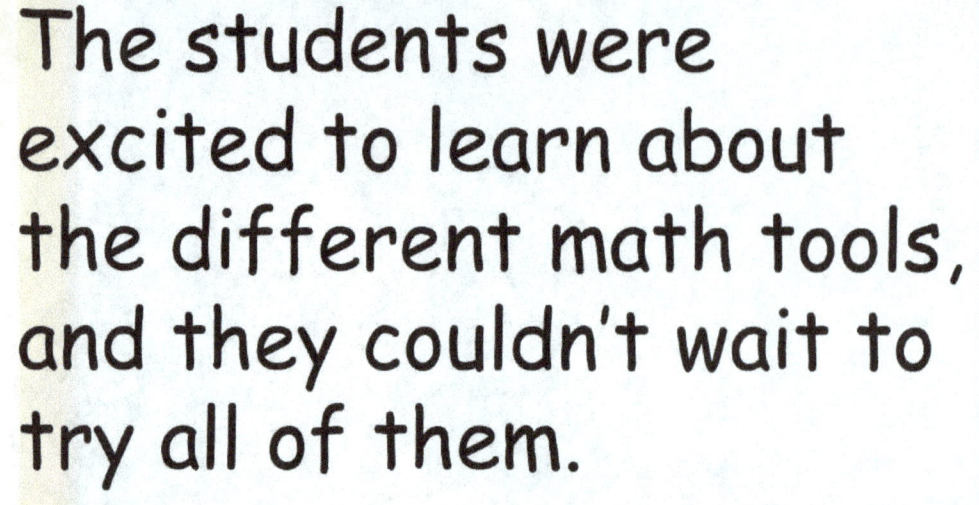

After Mr. Lee thanked the Super Solvers for visiting the school, they waved goodbye before leaving in the STEAM Machine bus.

SIGHT WORDS

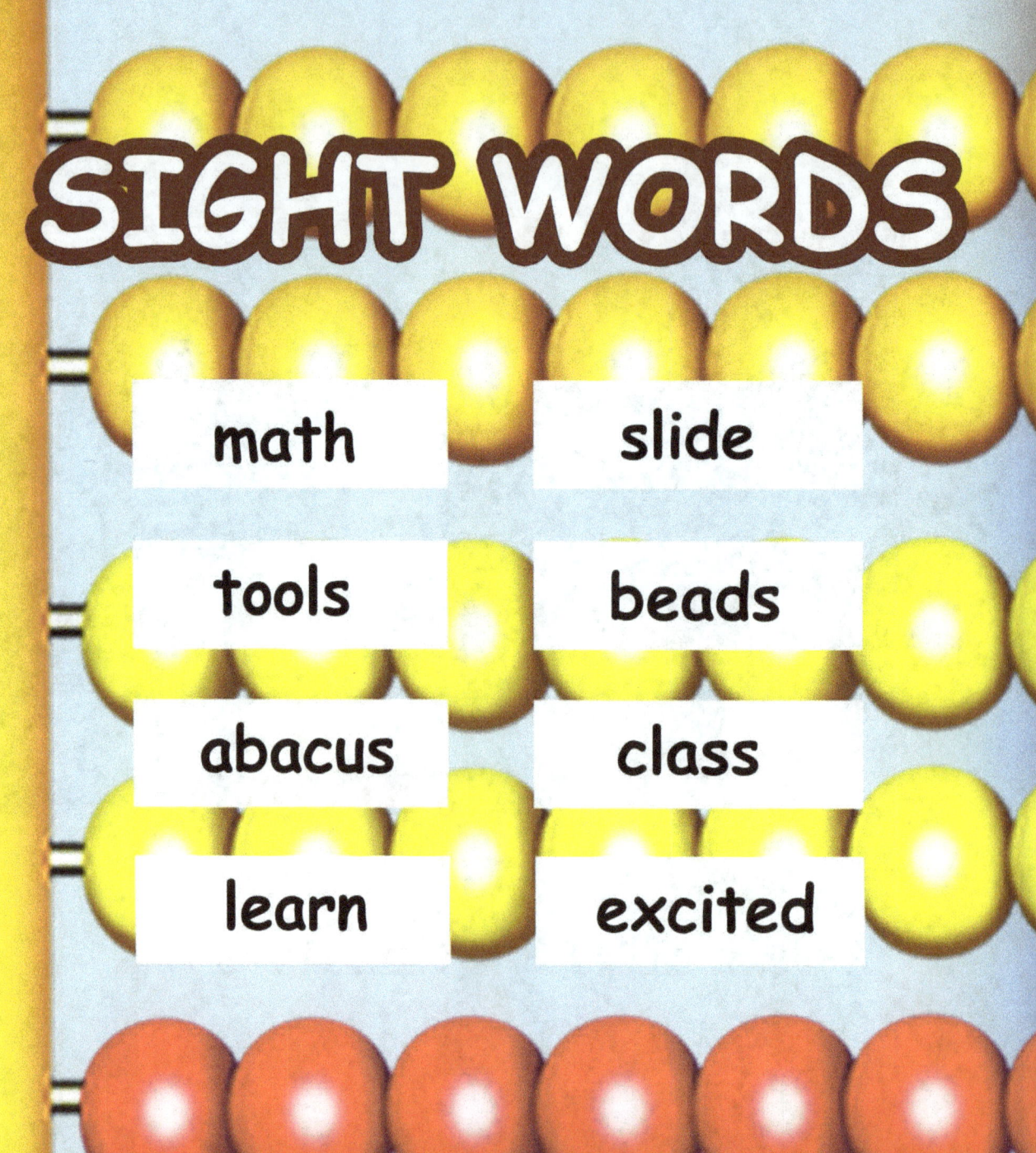

math

slide

tools

beads

abacus

class

learn

excited

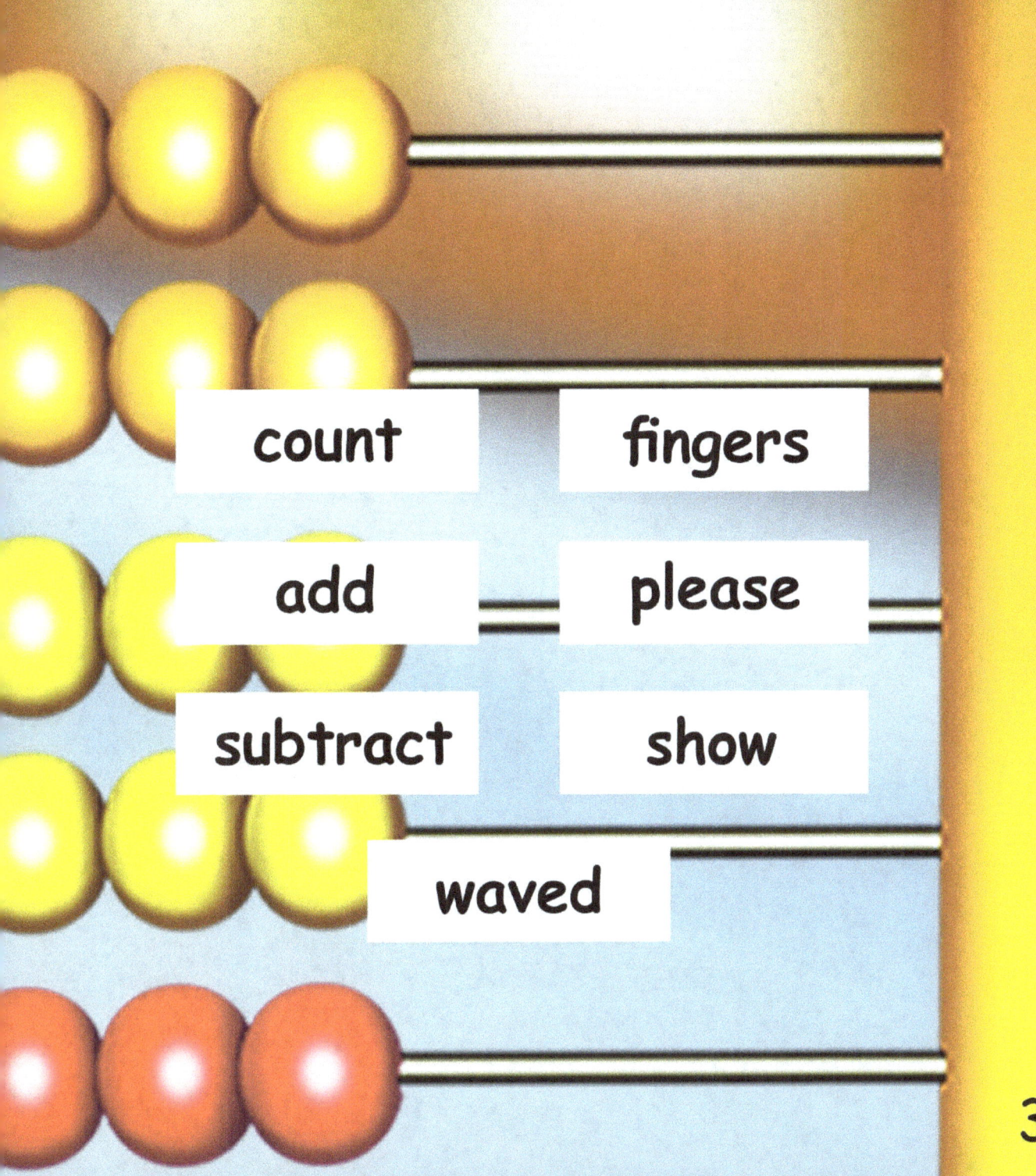

count   fingers

add   please

subtract   show

waved

About the Author

Dr. Maury Wills is a native of Austin, Texas. His skills as an educator have developed throughout a successful career in various fields of education. As an avid musician and drum major, he has always marched to his own beat, especially to that of a sound and quality education. For this reason, Dr. Wills has been an advocate for children, a supporter of educators, and collaborator for parents.

Pursuing his goal in education wasn't a small feat, but a great milestone, as he has served as a 1st and 2nd grade teacher, 5th grade teacher, special education teacher, and finally, becoming one of the youngest principals in his former school district at the age of 29. With his past experiences in the teaching and learning fields, community and parenting corroborative skills, Dr. Wills was sought after to lead the DeKalb Agriculture Technology and Environment Charter School, Inc.

His profound knowledge, tenacity, and dedication to his children have benefited the school community at large. His leadership and commitment to the charter movement have also increased continuous and ongoing academic achievement, better parental and community involvement, and a fiscal accountability that is recognized locally, regionally, and nationwide.

Dr. Wills' instructional leadership is deeply rooted in his philosophy of excellence and learning, "Every student will learn, given the opportunity to embrace a climate of high expectations, a committed support system, and connections to the context of the real world, the Three R's: Learning with Relationships, Relevance, and Rigor!"

The Adventures of the STEAM Machine

Did you enjoy this fifth book in the series?

Look for other STEAM books wherever books are sold!

<u>S</u>cience – Sprout, Bean, Sprout!

<u>T</u>echnology – Shine, Sun, Shine!

<u>E</u>ngineering – Go, Robot, Go!

<u>A</u>rts – Paint, Dance, Play!

<u>M</u>ath – Slide, Beads, Slide!